Apple Watch

Master Your Apple Watch

Complete User Guide From Beginners to Expert

Andrew McKinnon

Andrew McKinnon

Andrew McKinnon

<u>Disclaimer Notice:</u>

Please note the information contained within this document is for educational and entertainment purposes only. Every attempt has been made to provide accurate, up to date and reliable complete information. No warranties of any kind are expressed or implied. Readers acknowledge that the author is not engaging in the rendering of legal, financial, medical or professional advice.

By reading this document, the reader agrees that under no circumstances are we responsible for any losses, direct or indirect, which are incurred as a result of the use of information contained within this document, including, but not limited to, — errors, omissions, or inaccuracies

Table of Contents

Introduction

Apple is hailed as the powerhouse of consumer electronics design and user experience, but it is also an unexpected driver of modern fashion. When the iPod launched a decade ago, it was notable for storing and playing over 1,000 songs on a tiny device the size of a deck of cards. However, one of its lasting impacts was the introduction of white ear buds to almost every public space. Today, wearing them is so commonplace that we never even think about it. Yet, just ten years ago, they were a fashion statement screaming "I own an Apple product".

With the recent release of the Apple Watch, the company strives to make another fashion statement: "The wearable personal technology we have seen in movies can now be even more personal, more distinctive and more stylish".

The Apple Watch is not a sliver of glass you keep in your pocket or bag; It is essentially a timepiece with an integrated communications center, as well as a connection to your personal information and devices that you wear all the time, anywhere.

You may currently be wearing a wrist watch, or if you are like many people, you haven't worn one for years, since your last piece was replaced by the massive screen smartphone in your pocket. What is the appeal of the Apple watch as a timepiece?

The Apple Watch is essentially an extension of the data you have deal with every day, and a shortcut device to accessing it. Without making the distracting motion of getting your iPhone out of your pocket or dashing to the next room to get it, you can keep up with your schedule, notifications, and reminders conveniently with you at all times. You can also interact with people via phone or text messages, reach out to other Apple Watch owners via Digital Touch features, keep track of various fitness goals and record vigorous exercise information generated by your iPhone's and watch's sensors and safely and conveniently pay for your shopping using Apple Pay and much much more.

This eBook is your simplified and condensed guide to the Apple Watch and its possibilities and capabilities. If you already have the watch, you will agree with me that owning it is initially a delightful and surprising experience, and later a perplexing learning process as you discover its capabilities and learn how to solve problems with its help.

The buzz that Apple Watch has created in the tech world right now is hard to ignore. Cupertino's very first wearable gadget looks to become the next big chapter in the history of computing by bringing all your favorite apps, tools and notifications to your wrist. With all these capabilities in mind, we are bringing you this ebook to help you master the ins-and-outs of your Apple Watch, get to know what is inside and how you can use it to simplify your life.

Should you buy an Apple Watch?

Apple watch is a revolutionary new product with long established roots in technology and time-keeping. However, most people are confused as to whether spending over $350 accessory on a wrist device they don't really understand is worth it. If you purchased this eBook because you want to know more about the watch before you buy, or to help you decide whether to purchase one or not, you made the right move.

Andrew McKinnon

Everyone today needs a phone. Most people today need a computer. Much like the iPad, the Watch feels a lot like an extra accessory, which makes it difficult to figure out whether you need it or not. To make the decision process little bit easier, let's break down.

Whether you should buy an Apple Watch or not boils down to how compelling the main features are for your lifestyle – by themselves or when combined. These features include timekeeping, health and fitness tracking, informational and notifications widgets, remote control, Apple Pay and communication.

Simply put, the Apple Watch is the shuttlecraft to the iPhone's starship. Most of the activities the watch is capable of can already be done on your iPhone, the watch merely makes them more convenient to do. Only a select few are exclusive the Apple Watch.

Chapter 1: Getting Started

Set up and pair with iPhone

To connect your Apple Watch with an iPhone, a setup assistant will walk you through a few simple steps to pair the Watch with your iPhone and make it your own.

Place the Apple Watch on your wrist, then press and hold the side button until the Apple logo appears. When prompted, position the iPhone so that the Watch appears in the camera viewfinder on the iPhone's screen. Follow the steps on iPhone and Watch to complete setup. During setup, you will be prompted to choose your language, passcode, and watch orientation.

The Apple Watch app on iPhone

With the Apple Watch app on your iPhone, you can customize the watch options and settings and set up services such as Apple Pay for the watch. This application will give you access to the App Store where you can browse, download and install applications for the Apple Watch. To access these features, simply open the Apple Watch app by tapping on the Apple Watch app icon then tap My Watch to open the Apple Watch settings.

Power on and wake

To **turn on** the Apple Watch, press and hold the side button until you see the Apple logo. Note that a black screen may appear for a brief moment at first. The watch face will appear shortly after the Apple logo.

To **turn off** the watch, press and hold the side button until a slider appears on the screen. Drag the slider to the right.

To **wake** the Apple Watch, tap the display or raise your wrist. The watch will automatically **sleep** when you lower your wrist. Alternatively, if you are not wearing your watch, you can rake it by pressing the Digital Crown.

Chapter 2: Getting to Know the Apple Watch

For a brand new product in the market, Apple watch comes with numerous capabilities that are guaranteed to impress both the geeks and the noobs.

It would be easy to expect that the Apple Watch is like an iPhone you wear on your wrist, but really isn't the case. The watch specifically emphasizes short and focused interaction with your information. For instance, on the watch you can view new notifications and receive text messages, which you quickly send replies, but any lengthy correspondence will be shifted to the iPhone.

It's important to understand how this approach works. We are so used to devices that demand our full attention that it seems impractical at first that the Apple Watch does not demand the attention. You shouldn't be surprised if it takes you a couple of days to adopt this mindset when you finally get to use the watch, even when you know about it ahead of time.

The Watch and iPhone Connection

Although the Apple Watch does have its own touchscreen and features wireless communication capabilities, a microprocessor to run its systems and apps installed on its memory, the device heavily relies on a companion iPhone to be functional. The watch works only with iPhone 5, iPhone 5c, iPhone 5S, iPhone 6 and 6 Plus running iOS 8 or later. If you are a proud owner of an iPhone 4 or iPhone 4S the Apple watch probably isn't for you. Also, if you don't have an iPhone or feel that the extra features of the watch are not compelling, you may not find the Apple Watch as useful as the designers intended it to be.

The Apple Watch, has core functions such as the clock and activity tracker that are dedicated applications installed in the device's memory. However, because the device relies on the wireless Bluetooth or Wi-Fi communication with the iPhone for the most part, a connection with the iPhone is necessary. As a matter of fact, developers at the moment cannot write native 'Apple Watch Applications'. Instead, they create iOS apps that have the capability to display data on the watch's screen and be able to communicate with the watch in specific ways.

The Apple watch pulls the GPS positioning data from the iPhone and piggybacks on its Internet connection. This basically means that even if you are getting the watch for just lightweight task such as tracking your exercise activity, make payments via Apple Pay or just receive notifications when an important chat messages comes in, you would still need to carry the companion iPhone along to provide GPS data, calorie burn estimates during workouts and an internet connection to transmit data remotely.

The companion iPhone has an Apple Watch application that you can use to control the watch's Home app icon layout, settings such as message displays and fitness features and other applications. However, Apple is reportedly working on an update to the watch's operating system dubbed watchOS 2, the next version of the watch's operating system that will allow developers to create native apps that will be installed and run directly on the watch. These applications are said to come with full access to the watch's sensors, display and other hardware.

Charging the Apple Watch

According to many users of the Apple Watch, Apple's estimate of the battery lasting for up to 18 hours of normal use is conservative, but you should expect to charge your watch once daily. Exercise tracking drains the battery the fastest largely because the heart beat sensors in the watch are powered locally by the device's battery. At the end of the day, expect to charge the phone in readiness for the next day.

The watch automatically switches to Power Reserve mode when the battery hits 10 percent. This is when everything except the basic digital readout of the time is switched off. If you notice the battery drains fast, it may be wise to manually switch the watch to Power Reserve mode whenever you are not actively using it.

The Apple Watch comes with a magnetic charger that attaches to the back of the watch to recharge the battery. It uses the inductive charging system which makes it easy to connect to charge even in the dark.

Apple Pay and Passbook

Other than Bluetooth and Wi-Fi, the Apple Watch comes with another wireless technology that enables the user to make payments through Apple Pay. NFC (near field communication) is a radio communication technology that works with NFC-based payment readers, but like other services, only works when the watch is paired to your iPhone.

With Apple Pay, instead of swiping your credit or debit card to make purchases at a store, all you have to do is hold out your wrist close to the NFC reader and double-press the side button on the watch. Apple Pay will transmit your payment information using an encrypted token; the transaction will therefore be safe, convenient and fast.

When you enter your card details to enable Apple pay, the details will appear on the Passbook application on the watch. Other payment related information such as royalty cards and tickets that create scannable bar codes are also stored here. You can use the Passbook to make quick payments at the train station, to buy coffee at Starbucks, get admission to a movie theatre and even present your boarding pass at the airport just by scanning your watch. More details on this feature are covered in the coming chapters.

Security

Today, one of the greatest concerns when it comes to new technologies is how safe they are. With the Apple Watch, data security is enabled with the use of a passcode, just like with the iPhone security. As a matter of fact, to use Apple pay, you need to set a Passcode. Turning the passcode on your watch off automatically removes your Apple Pay cards from the system and will require you to set them up again when you need to use them.

When your watch has an active passcode set, it will automatically lock when your remove it from your wrist. The idea here is to prevent thieves from accessing Apple Pay and other services or data if they steal the watch. It is expected that watchOS 2 will have additional security features such as Activation Lock that renders an Apple Watch completely inoperable without an Apple ID and password.

Chapter 3: Choosing the Right Apple Watch

There are quite a number of variants and models of the Apple Watch to choose from based on core specifications, colors and strap types. Choosing an Apple Watch is nothing like choosing an iPhone or a MacBook – forget about selecting one based on storage. Under the hood, the Apple Watch hardware and functionalities are identical across the board, you will get to select one by looking at the alloys – stainless steel, aluminum and Gold.

The most important factor to consider when choosing the watch is how it will fit your wrist. Most of the bands come in different sizes but some are limited to only one case size, so you should know what size your wrist is before settling on the band of watch to go with. The three models of the Apple Watch are Apple Watch (classic), Apple Watch Sport and the Apple Watch Edition. All the three models of the Apple Watch come in two different sizes and several bands.

Apple Watch

The defining feature of the singularly named Apple Watch model features a stainless steel case, available in regular shiny stainless steel and dark space black tint. It also features a sapphire crystal surface providing scratch-resistant protection to the display.

Apple Watch Sport

The Apple Watch Sport is specially designed for exercising, hence the lightweight anodized aluminum case. This material is a custom alloy that is up to sixty percent stronger than ordinary aluminum. The watch also features an Ion-X glass display, a glass variant that is more scratch-resistant and stronger than ordinary glass but much lighter. However, this glass is less protective than the sapphire glass. One watch variant comes in a Space Gray finish and a black band.

Apple Watch Edition

If you are a flush buyer looking to make a statement with your Apple Watch, the Apple Watch Edition is for you. This model of the watch features an 18 karat gold case in rose or yellow hues and a sapphire crystal cover. Apple contracted its own metallurgists to formulate their own gold case that is twice as strong as standard gold. Because gold is a fairly soft material, high end watchmakers often hire specialists to develop custom mixes that are stronger and less susceptible to dents.

Model Sizes

Whether you have tiny hands or gigantic wrists, rejoice because there is an Apple Watch that fits your size. For each model of the watch, there are the 42mm and the 38mm variants to choose from.

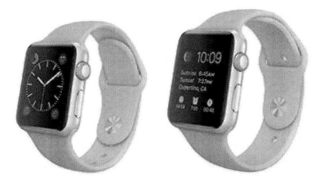

Before making a decision on what size of the watch to buy, you can actually download a sizing guide from Apple store or preview an actual-size render of the watches on your iPhone or iPad using the Apple Store app. Simply tap on Apple Watch to access options then select the watch model and price to get a link to Compare Case Sizes.

Actual size

38MM	42MM

Selecting Watch Bands

The Apple Watch offers six different types of bands to secure it to your wrist. These interchangeable bands range from the luxurious Milanese Loupe to the sweat and water resistant fluoroelastomer bands in the sport variants. The bands are visually stunning and are thus best browsed and selected visually. However, there are a few takeaways you will useful:

- The bands fit the watch by sliding into notches at the top and bottom of the watch. If you prefer, you can have more than one band that you can easily change depending on the activity – a leather classic buckle when heading out to the office or the sport band when going for your morning run.
- The bands are designed to be easy to adjust. Most bands come with magnetic embedded in the strap to make it easy to snap into place and to find a comfortable diameter to fit your wrist. The stainless steel bands have buttons that makes it easy to remove or add the links without any special tools or taking it to a technician.
- The Apple Watch Edition models are available with matching connectors and gold buckles.
- Apple has released specifications of the watch's bands for manufacturers to design and create third-party bands. Soon, there will be a world of variety flooding the market and you can find them at very pocket-friendly prices.

Apple currently sells additional bands for the watches separately online in the Apple Store as well as via the Apple Store app for iOS.

Apple Watch Pricing

You can purchase the Apple Watch online at the Apple Store starting from $349 for the 38mm Apple Watch and $549 for the

Apple Watch classic. For each of these variants, the 42mm version of the watch costs $50 extra.

The Apple Watch Edition models are only available at 'Select retail stores' and they retail from $10,000 all the way to $17,000. Initial supplies for this model have been limited

Andrew McKinnon

Chapter 4: Interacting with the Apple Watch

When Steve Jobs launched the iPhone back in the year 2000, he reiterated that the ideal tool to use to interact with it was not physical buttons or a stylus, but our fingers. The touchscreen responsiveness made finger gestures —tapping, pinching, and swiping — the newest language of how users interact with technology. However, when it comes to small display devices like the watch, fingers are usually too big to be considered effective or practical for that matter. To overcome this limitation, Apple has incorporated additional interaction methods in the Apple Watch.

Gestures

To use the Apple Watch and its apps, there gestures you need to know are Tap, Firm Press, Swipe and Drag. The watch display responds to these touch-based gestures as well as arm movements.

Swipe   **Drag**   **Tap**

Press firmly

The Digital Crown

Let us begin with the signature new controller on the Apple Watch, the knob on its side called the Digital Crown. A crown has been a staple of mechanical watch design for centuries, used to set the time and, on some models, to wind the mechanism that keeps the time coil running.

What is "digital" about the Digital Crown, you may ask. The crown does not operate using mechanical gears like the traditional crown; it instead translates rotary movements into digital data by use of internal infrared LED lights and photo diodes

Turning the Digital Crown on the Apple Watch scrolls content on the screen, switches between visible options and zooms in and out when viewing photos or maps and when on the Home screen.

The Digital Crown also has button that when pressed, produces various effects, depending on the context:

- When the watch face is visible, pressing the Digital Crown once displays the Home screen.
- Pressing the Digital Crown again returns to the watch face.
- While viewing a glance or notification, pressing the crown takes you back to the watch face.
- Pressing and holding the crown for one second launches the Siri's voice-activated interface.
- When you are on the Home screen and any application other than the face is centered, pressing the Digital Crown centers the face; pressing it again reveals the face itself.

- While on the Home screen, you can rotate the Digital Crown to launch and zoom in to the centered app (the default app is the watch face).
- Double-pressing the Digital Crown switches between the watch face and the last app you used.

The Side Button

The Side Button is the name of the physical button just below the Digital Crown. Pressing the Side Button on the Apple Watch brings up the circle of friends (12 contacts to be specific) so you can initiate communication i.e. voice calls, text messages or Digital Touch. Double pressing the Side Button launches the Apple Pay service.

Interacting with the Screen

The Apple Watch features a touchscreen Retina display, you can therefore tap, pinch or drag as you can your iPhone screen. In fact, when you get used to the Apple Watch display, you will find that scrolling with a finger instead of using the Digital Crown is more convenient.

Apple has added watch-specific views called glances to the Apple Watch interface. You can swipe from the bottom up on the watch face to show the calendar, the day's events on the calendar or show music playback controls etc. You can also swipe left or right to switch between various watch glances.

Unlike the touchscreens of other iOS devices which don't matter how hard you tap, the watch can tell the difference between a normal gentle tap, a firmer force touch and a long press. The screen registers pressure sensitivity when tapped or pressed.

Siri on Your Wrist

The Apple Watch does not have a virtual keyboard, meaning that you cannot enter text on its touchscreen. However, Siri, Apple's popular voice-activated intelligent assistant, can be launched by raising the watch and using the phrase, "Hey Siri," or simply pressing and holding the Digital Crown.

Just like Siri on the iPhone and iPad, you can request info, ask for directions, dictate messages, view and create calendar events, and more. You can do much of what you can do with Siri on the iPhone because the watch actually just relays the stream of data from the Internet via the iPhone.

Taptic Feedback

You don't always need a screen or a button to interact with the Apple Watch. The watch has a Taptic Engine, a linear actuator that taps your wrist as a way to provide feedback when something happens e.g. when you receive a notification.

Unlike the iPhone's muted vibration that is so strong it can shake a table, the watch's Taptic Engine is subtle. This feedback is contextual, as well: for instance, when you are following a route in the Maps app, distinctive taps will let you know whether an upcoming turn is a left or right.

The Taptic Engine combined with the speaker driver creates a tactile feedback for actions such as firm-pressing and force-touching the watch's screen.

The Taptic Engine feedback is also great for those moments when you don't want to miss calls or messages because you don't feel the iPhone's vibration in their pocket, or is set to silent when you are in a meeting. A small tap and quick glance at your watch face identifies the message sender caller.

Chapter 5: Personalizing the Apple Watch Face

Considering that the Apple Watch is as much a fashion statement gadget as it is a piece of functional technology, personalizing it goes beyond mere selection of the case material and type of watch band. Its Retina screen can display analog, digital, and even several whimsical watch faces.

Picking a Watch Face

The Apple Watch offers a number of watch faces - from simulated traditional chronographs to Mickey Mouse with tapping toes to an astronomical solar system design.

To customize the watch face, force-touch the screen when the time is shown to show other faces saved in memory. Swipe left or right to browse then highlight the one you want and tap to select it.

Customizing the Face

Some faces include elements that you can add or remove, while others offer simple options such as changing the colors.

Step 1: Force-touch the face then tap the Customize button when it appears.

Step 2: Tap on an outlined interface item to choose it and use the Digital Crown to scroll through the available options.

Some faces offer multiple customization options. For instance, in the figure, swiping left highlights the whole interface (take note of the indicator dots at the top of the screen; the dot on the left is brighter), and rotating the crown varies the color of the interface.

Step 3: Force-touch the watch's screen to return to the face selector then tap the face to activate it.

Using Face Elements

Face elements (also called complications) on some faces furnish the user with quick information, such as current temperature and message count etc. but they are not static. To use a complication, simply tap on it to initialize related app. For instance, tapping the temperature complication launches the Weather app, dates or listed events icons open the Calendar app and the activity circles open the Activity app.

The Chronograph face has an included stopwatch that is always available for immediate use, so you wouldn't need to launch the Stopwatch app when you need it.

- Tap on the Stopwatch button at the top-right corner to start recording time.
- Tap on the Lap button at the bottom-right corner to track a second lap time; you can tap it again for additional splits.
- Tap on the red button to stop recording time.
- Tap on the lap button when no longer recording to clear the stopwatch results.

Saving Custom Faces

With the variety of watch complications available, you can furthermore save combinations for later use.

The Modular face, for instance, packs lots of information on a single screen, but perhaps on a weekend you would prefer to see the sunrise and sunset information in the middle of the face, and instead of the battery indicator, have the stopwatch or a countdown timer. You could even want to save a few different colors to make switching easy and fast. All these are possible because you can save multiple customized faces to swipe and activate them easily at any time.

Glances

Although the Apple Watch runs apps, often, you wouldn't want to launch an app just to view a brief snippet of information. This is where glances come in handy. A glance acts as quick window to data and offers several convenient controls, a lot like Control Center on the iPhone. To understand them better, think of glances as a row of screens that hide below the watch face that can be easily accessed by swiping them up.

To browse through various glances, swipe from the bottom of the watch face up to reveal a glance then swipe left or right to browse the installed glances. The dots at the bottom of the screen show which glance you are viewing and how many glances are to the right or left of the screen.

When you tap on anywhere on the active glance, the associated app will initialize. Most glances are interactive, for instance, the Settings glance features controls to enable Do Not Disturb and Airplane modes, Mute Audio and locate your phone among others. The Heart Rate glance shows your heartbeat and the Now Playing glance brings up remote controls for multimedia i.e. next, pause etc.

How to Install and Organize Glances

Go to the Apple Watch app on your iPhone and tap on Glances. This will take you to the Glances screen.

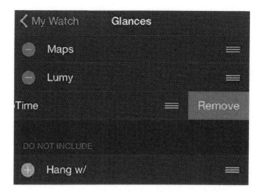

- To remove a glance, tap on the minus sign button to the left of the glance and then tap Remove.

- To add a glance, tap on the glance's green plus button below the Do Not Include section.

- To adjust the order of the glances, simply drag the handle icon to the right of a glance up or down.

You can also add or remove glances right from the app's settings in the Apple Watch app, through the main My Watch screen. The watch allows you to set up to 20 glances, but this would be a nightmare having to scroll back and forth to find a specific glance on the watch. With glances, fewer are always better.

Andrew McKinnon

Chapter 6: Apple Watch Apps

Developers who build apps for the iPhone can build versions for the Apple watch. Remember that the watch essentially communicates with the iPhone to retrieve app data. As a matter of fact, you may be surprised to find that most apps you already have on your iPhone now have Apple Watch support components.

Locating and Opening Apps

Unless you are viewing glances or notifications, pressing the Digital Crown on the Apple Watch displays the Home screen like the one shown below.

To view apps that are presently outside the edges of the screen, slide your finger on the display. You can also set the Digital Crown to scroll through the installed apps.

- To open an app, tap on its icon.
- You can also turn the Digital Crown to zoom out to reveal all apps, or to zoom in and open whichever app is centered.
- You can also open apps the hands-off way: Just raise the watch and ask Siri to open the app. Say "Hey Siri, open then app's name."

Customizing the Home Screen

App icons are ordered from the center outwards, with the watch icon always taking the center spot. You can arrange the app icons whichever way you want, in two ways:

- The first is by touching and holding any app icon until all the icons are gently vibrating and the highlighted app icon appearing larger. You can then move the icon to a new location. Once placed where you want it, you can move another app icon simply by touching and holding it until it is selected then move it. Note that this does not require a force or firm touch; just give the icon a gentle touch wait a second.
- Alternatively, you can customize the watch's Home screen from the Apple Watch app on your iPhone. Open the Watch app and navigate to App Layout. Touch and hold an icon to select it, then drag it to a new position to place it there. This is by far the easier way to organize apps on the watch's home screen.

Install Apps

To choose which apps to install on the watch, follow these steps:

Step 1: Open the Apple Watch app on your iPhone and scroll all the way down past Apple's built-in apps. Any iPhone app that has a watch component will appear here.

Step 2: Tap the on the name of the app you wish to install.

Step 3: Tap the on Show App on Apple Watch switch to the right of the app to turn it on. The app will appear on the watch's Home screen almost instantly.

If you wish to show an app on the watch's glance screen, tap the app's name on the main screen (My Watch) on the Apple Watch app on your iPhone to access the app's detail screen then tap on Show in Glances option to enable it.

Remove Apps and Glances

If you no longer want an application on the Apple Watch, removing it is easy.

Step 1: On your watch, go to the Home Screen and touch then hold the icon to make the icons vibrate.

Step 2: Tap on the app icon you wish to remove and confirm the action by tapping on the Delete App button. You can also remove a third party app by tapping on the X at the top left of the app icon when the icons are gently vibrating.

Step 3: A quicker way to manage your apps on the iPhone is to open the Apple Watch app on the phone, then tap on the app's name and tap on the Show App on Apple Watch switch to turn it off. Note that when you remove an app from the watch, the app's installed glances will also be removed.

Find New Apps

Your Apple Watch apps are tied to iPhone apps. Apple, in deciding on this formula, has come up with a winning strategy to distribute apps - the App Store. To discover what apps have Apple

Watch components, open the Apple Watch app on your iPhone then tap the Featured button at the bottom of the screen get a list of apps with watch components.

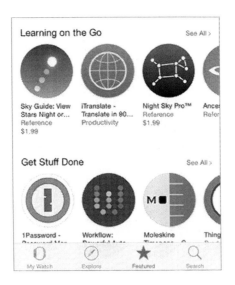

When you purchase an app from the App Store, it downloads to the iPhone, and if it has a component, it appears in the Apple Watch app so you can install on the watch itself.

With watchOS 2 expected at the end of 2015, the Apple Watch is expected to support native apps, meaning that you will be able to directly download and install apps exclusively for the watch without necessarily having to install them on the iPhone.

Chapter 7: Notifications

Of the many features the Apple Watch boasts of, users will take advantage of notifications the most. When an incoming call, text message, bank card alert, Twitter reply, or other notification arrives, the watch displays it. As a matter of fact, with the iPhone, asleep in my pocket, it doesn't even chime or vibrate. (However, when you're actively using the iPhone, notifications will not appear on the watch.) Notifications are the heart of the watch's interaction. It has made checking notifications on your wrist quick and discreet when needed, without the need to pull out the iPhone.

Acting on Notifications

When there is a new notification, the watch gently taps your wrist and shows the notification on the display. Depending on the type of notification, there are several actions you can take.

- Tap the app icon or title of the app— if it exists on the watch - to open it. If you tap on a notification icon of an app that does not exist on the watch, nothing will happen.
- You can tap the Dismiss button to acknowledge a notification you have read but need not open the app. Also, swiping notifications from the top of the screen down of the screen dismisses active notification.
- Depending on the app, tapping on the action button has a special response or action. For instance, tapping on the Instagram notification opens the image post and tapping on a new text message notification is a shortcut to send a quick reply.

- Notifications disappear after a few seconds of inaction when the watch is in use. Alternatively, you can flick your wrist to put the watch screen to sleep and dismiss active notifications. Dismissing notifications adds them to the Notification Centre.

Keeping Notifications Private

If you want to make sure that no one can see watch notifications, for instance text messages as messages arrive, navigate to the Apple Watch app on your iPhone, tap Notifications, then turn on the Notification Privacy switch.

This setting ensures that your notifications are kept private. When there is a new incoming notification, only the associated app and the sender will be displayed. You will need. Tap on the notification to view its details.

Viewing Missed Notifications

When you have unread notifications, there will be a red dot at the top of the watch's face. To view hidden notifications, simply swipe down to access Notification Center to see hidden, ignored or missed notifications. Use the swipe action or the Digital Crown to scroll through the list.

Clear Notifications

If you have many notifications piled up in the Notification Centre and want to clear them up, open the Notification Centre and swipe left to reveal the Clear button. Tapping on this button will remove the notification.

Force-touching the Notification Centre screen offers the option to Clear All notifications. This gets rid of all the notifications.

Customize Notifications to Receive

The downside to notifications is the likelihood of being overwhelmed by too many of them — you wouldn't want the Taptic Engine to feel like massager on your wrist. This wouldn't be good for your watch's battery. By customizing some settings in the Apple Watch app on your iPhone, you can take control of which notifications get pushed to the watch.

In the Apple Watch app, tap Notifications then scroll down to the list of apps. To customize which application notifications get pushed to the watch, tap Mirror my iPhone button to turn them on or off. In general, the watch mirrors the iPhone's notifications.

For some apps such as the Calendar, you can customize specific calendar items that get pushed to the Watch. Simply tap on Calendar > Custom then tap on the type of notification to turn it on or off.

For third party applications, you can only turn their notifications On or Off.

Setting Feedback for Alerts

Notifications on your Apple Watch can alert you by chiming or by tapping on your wrist. You can select choose whether to be alerted by Haptic feedback or chime only on Apple's in-built apps. This option is not available for third-party applications.

To customize application notification style, tap on Notifications on the Apple Watch app on your iPhone then tap on the app you wish to customize e.g. Messages. Tap Custom and select the alert styles for the app.

Receive Mail from VIPs Only

Do you wish to be alerted only when you receive new messages from a special list of people? First, you must mark the people you want to receive mail notifications from as VIP.

Step 1: Open the Mail app on the iPhone and locate an email message from the person you wish to add to the VIP list.

Step 2: Tap on the person's name to open their contact details.

Step 3: Scroll down the contact details and tap Add to VIP.

Next, change your notifications to receive notifications when contacts on the VIP list send you an email.

Step 1: Open the Apple Watch app on your iPhone and go to the Notifications section.

Step 2: Tap on the Mail icon and choose Custom.

Step 3: Set the Show Alerts from VIPs slider to On.

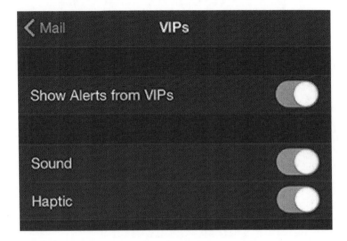

When you receive a new message form anyone you have added to the VIP list, a notification will appear or the message will be shown as a new notification on the watch

Chapter 8: Communicating with Friends

Your iPhone is a communication device – that is easy to understand. The iPhone is designed to enable us to make calls, send text messages and communicate through video calls and conferences. The expanded capabilities of the iPhone also enable us to participate in social networks like Facebook, Instagram and Twitter. The question is, how does the Apple Watch fit in this equation? When adequately utilized, the Apple Watch is just as great a communication device as the iPhone, as it incorporates text and audio messages through the Messages app and even makes and receives phone calls. The watch also brings a new form of communication - the Digital Touch.

Communicating Using Messages

When you receive a text message on your iPhone, a notification will be pushed to your watch, which will notify you via the Taptic Engine. To reply to a new message, you can:

- Raise your wrist to view the message and the sender information.
- Tap Reply and send a quick reply to the message. You can enter the reply in three different ways: by selecting a preset text; choosing an emoji image; or dictating a response.

Reply by selecting a preset text

Because the Apple Watch does not have a keyboard, entering text the traditional way is out of the question. However, Apple has come up with an ingenious way to save text message replies in 'quick boards' that you can quickly select to send. The quick boards feature analyzes the received text message and presents the most likely preset replies based on the contents of the original message. Alternatively, you can scroll down to select a different message from other preset replies and tap on one to select and send.

Although Apple's preset replies try to be useful in many situations, they are pretty generic. However, you can create your own custom preset replies on the iPhone. To do this, go to the Messages section of the Apple Watch app on your iPhone and select Default Replies. Enter the text you want to appear on the preset list (including emoji) and save.

Reply via Emoji

To send a quick emoji reply, tap on the emoji button at the bottom left of the screen then swipe sideways to select an emoji from the various animated faces on the list. There are several large animated emojis to choose from including hand gestures and hearts and many smaller ones. You can change the variation of the emojis including animation using the Digital Crown.

Select the appropriate emoji then tap Send to dispatch the emoji reply.

Dictate reply

To dictate a message, tap the microphone button and speak your message reply. If your message includes punctuation in the text area of the message, be sure to dictate it as well. e.g. say "What time will you get here question mark" to send "What time will you get here?"

When done, tap Done to continue. Choose the mode in which the message will be sent. Select Send as Text to reply as a text message or Send as Audio to send it in a waveform. Your message will then be sent.

Create a New Message

You will need to activate Siri to create and send a new text message. Do this by pressing and holding the Digital Crown or by lifting your wrist and uttering the phrase "Hey Siri". When Siri is running, simply say "Text [contact name]" followed by the message you want to send. Alternatively, you can activate Siri then wait for it to prompt you to speak the message once you have selected the recipient from your contacts. To send the message, invoke Siri again and say "Send".

Another way to create and send a new message is to press the side button on the Apple Watch to bring up the contacts list, tap on the icon of the person to send the message to then tap the Messages button located on the bottom right corner to create a new message.

There is another way to send new messages and send replies to existing messages from your Apple Watch. Simply open the Messages app by tapping on the icon on the Home screen, choose an existing message thread and tap on reply at the bottom of the screen. To create a new conversation, Force-touch the messages screen to bring up new options then tap on the New Messages button to create a new message. Add a recipient by tapping Add Contact then tapping the contact from the list of frequently contacted people. If the recipient is not on the list, tap on the Contacts button at the bottom of the list to view your entire contacts list or tap on the Microphone button to dictate the name of the recipient.

Access more options on Force-touch menu

While viewing a conversation in the Messages app, you can force-touch the screen to reveal more messaging options:

- A Details button shows the contact information for the message sender or recipient.

- A Send Location button sends your current position in a virtual card that the recipient can open in the Maps app on the watch, iPhone, iPad or Mac.

- A Reply button allows you to send back a quick reply. You can also find the reply button at the bottom of the conversation or message.

Using Digital Touch

The Digital Touch feature of the Apple Watch is a swift and whimsical way to communicate with another person else who owns an Apple Watch. To initialize the Digital Touch, press the side button to see the Friends screen then tap a contact or rotate the crown to scroll through the friends list and tap the middle of the screen (or just wait for a second for the selected contact to open automatically).

If the contact owns an Apple Watch, you will see a Digital Touch 🔘 button at the bottom of the screen. Note that this feature only works between Apple Watches.

- Tap the Digital Touch button or on the middle of the contact's icon to open the Digital Touch interface, which will initially be a blank screen.
- Tap the screen to send a short Taptic Engine tap, or use your finger to draw something on the screen; the image you draw will be sent to the other person. You can choose different colors to draw with by tapping the colored circle at top right of the screen.

You can also send your heartbeat by simply placing two fingers on the Digital Touch screen. Romantic, isn't it?

When you receive a Digital Touch communication, it will show in the notification as a message. You can view the animation by tapping the replay button on the top-right corner of the Digital Touch screen.

Edit the Friends Screen

By default, the Apple Watch populates the Friends screen by selecting a dozen contacts that you have marked as favorites on the iPhone Phone app. However, you can change who appears on the Apple Watch friends list via the Apple Watch app on the iPhone. Simply open the app on the iPhone and go to Friends. Here, you can :

Remove a contact by swiping right to left then tapping the Remove button that appears; or by tapping the Edit button, then the red minus-sign button next to the person, and then tap Remove.

Add a contact by tapping in any empty slot then selecting Add Friend then choosing a contact from your contacts list.

Rearrange friends by moving their entries in the circle. A friend's position is indicated by the icon to the left of the entry. To rearrange the list, tap the Edit button then drag the handle icon ▤ to the right of the person's name. When finished, tap Done.

Note: *Because you can initiate a Digital Touch from the Friends screen, if you want to communicate with a contact via Digital Touch you must add them to your Friends screen.*

Communicate by Phone

You can receive and make phone calls right from your Apple Watch, and yes, I am sure it will look strange talking to your watch and it talking back to you. Making calls from a watch is not as private as when using a call, but like the many features of the watch, the phone calls capability is a great tool for brief calls or when it is not appropriate or convenient to use your phone.

Receiving a Phone Call

When you have an incoming phone call, the watch offers you several options:

- **Answer the call** on the watch by tapping on the green Answer button.
- **Answer on the iPhone** by turning the Digital Crown and tapping the option to answer on the phone.
- **Reject the call and send a quick text** by turning the Digital Crown, and tapping the option to send the text.
- **Mute the ringer** by holding your palm over the watch face for three seconds.

- **Reject the call** and instead send the caller to voicemail by tapping the red button.

- When you have a voicemail, it will appear as a notification and give you the option to either listen to it then on the watch or listen to it later using the Phone app on the watch.

While you are on a phone call on the watch, you can change the volume setting using the Digital Crown or by tapping the Mute button to silence the speaker.

Starting a Phone Call

- **Ask Siri to start a call**: Raise the watch and say, "Hey Siri, call [*contact name*]" (or activate Siri first by pressing and holding the Digital Crown and when prompted, say the contact name).

- **Start a call on the Friends screen**: Press the side button, select a friend to call, then tap the Phone button on the lower-left corner.

- **Start a call from the Phone app**: Open the Phone app and select a category from Favorites (from your iPhone), Recents, from the Contacts list or from Voicemail. Tap on a contact or recent call to make an outgoing call.

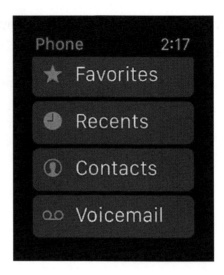

Voicemail

Managing voicemail on your Apple Watch is a lot more convenient than having to dig out your iPhone. From the main screen of the Phone app on your watch, tap Voicemail, then tap a message. You can play the voicemail message, return the call by tapping the Phone ◣ icon or delete the message by tapping the Trash ▉ icon.

Note: To switch to the iPhone during a call, simply drag up on the phone Handoff icon on the bottom left of the iPhone's lock screen.

Communicate Using Mail

The Apple Watch email ties into the Mail app on the iPhone primarily to enable you read and sort your Inbox. There is currently no way to reply to messages or compose new ones on your watch though, you would need to perform these actions on the iPhone.

Reading Email

To read mail on your watch, open the Mail app from the watch Home screen to see your Inbox and a couple of last email messages. Despite such a small space, the Mail app delivers quite a decent amount of information.

A blue dot on a message indicates that the message is unread, while icons on the lower-right corner indicate messages from a VIP contact, whether a message contains attachments, or has been replied to or forwarded.

A double angled bracket character (») on the top right corner of a message shows that it is part of a thread.

- To read a message, tap on it.
- When a message is open, you can access additional options by force-touching the screen. The options include Flag to flag the message, Unread to mark it as unread and trash to delete it and send it to trash.

You can access a variety of message options right from the watch inbox by swiping the message left. From here, you can tap Trash or tap More to access the Flag and Mark as read options.

Writing an Email

To compose a new message send a reply to a message, you will have to switch to your iPhone and drag up the Mail Handoff icon on the lock screen.

Personalizing Mail Browsing

While you cannot send mail directly from your Apple Watch, at least you can change some of the features on how you view it.

Open the Apple Watch app on your iPhone, go to Mail, then tap Custom, and have the option to make these adjustments:

If you want to switch to the iPhone during a call, drag up on the phone Handoff icon that appears on the iPhone's lock screen.

- The Inbox is the default mailbox displayed, but you can select from several other options, including other mailboxes, VIPs and flagged or unread messages. You can also choose to view only those messages where your email address appears in the To or CC field.

- **Message Preview**: Choose to view one or two lines of text, or select None to see only the basic information including the sender, subject line, and time of message.

- **Flag Style:** If a message is flagged, an orange dot will appear in front of the sender. Choose Shape to see a small flag icon in the lower-right corner instead.

- **Ask Before Deleting**: When you switch this option on, you will need to confirm every time you delete a message.

- **Organize By Thread**: Group your messages in a thread together (turn on) or list each message separately based on timestamps (turn off).

Chapter 9: Maps and Directions

Instead of staring at your phone while following directions from the Maps app on your iPhone, the Apple Watch allows you look at your wrist for directions instead. Isn't that easier, better and safer? Thanks to taptic feedback, you don't even need to look at your wrist at all.

The Maps app on the Apple Watch works as a pair with the Maps app on the phone. To To pan across the map, drag with one finger, and to zoom turn the Digital Crown when the map is running. Also double-tapping on the map screen zooms and centers on where you tap. To quickly find your location on the map and focus the map on your location, tap the Tracking button on the lower left corner of the screen if your position is not already centered.

Find a Location Using Siri

Step 1: Raise the watch and call up Siri using the phrase "Hey Siri". Alternatively, you can start Siri by pressing and holding the Digital Crown.

Step 2: Ask Siri to find a location: You can as a specific landmark or business, the location of one of your contacts, or a question like, "Where is the best coffee near here?"

Step 3: From the results presented, tap a location to see more details about the location.

Step 4: Scroll down to find the location address on the map, then tap it to open it on the Maps app.

Find a Location Using the Maps App

Using the search feature of the watch's Maps app gives you access to recent and favorite locations, besides enabling searching using dictation:

Step 1: Force-touch the screen while on the Maps app then tap Search.

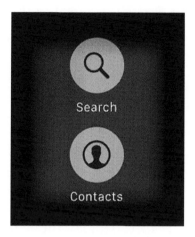

Step 2: On the next screen, tap on the Dictation (microphone) button to speak a search query, just as you would do after you invoke Siri.

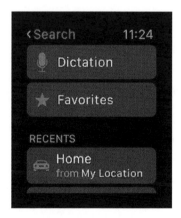

Alternatively, an easier way to access previously searched locations is by tapping the Favorites button to jump to saved destinations. You can also scroll down to the Recents list, which is a list of stored addresses and prior directions, as well.

Step 3: Whichever method you prefer, tap the destination on the presented results to view it on the map.

To locate a person or business in your contacts list, Force-touch the screen then tap the Contacts button. You can then to scroll between letters of the alphabet by turning the Digital Crown faster rather than scrolling single contacts at a time. This is particularly useful when want to reach your friends all the way down the alphabetical order.

Scroll through the contacts list until you find the person or business to locate and then tap the desired entry. Next, tap the address presented to locate it on the map.

Place a Pin

Do you need to mark a spot on the map? Perhaps to remind you where you parked your car, or to mark a general area for later? Simply place a pin by first touching and holding the spot on the map for a few seconds and a purple pin will appears. You should not force touch on this step, simply rest your finger on the screen.

When you tap on a set pin, it will bring up any information associated with the location such as the name of a business or the address. That screen also offers the option to Remove Pin button for when you want to get rid of the pin.

While you cannot move a pin once placed, you can drop a new pin in any other location on the map. However, you can only drop a single pin is at a time.

Adding Items to your Favorites List

Step 1: Open the Maps app on the iPhone, and find a destination by carrying out a search on the Search field at the top of the phone's screen.

Step 2: When the location's pin is selected, tap on the Share button.

Step 3: Tap on the Add to Favorites button in the share sheet that appears.

Step 4: You have the option to rename the new location. Next, tap Save to add to favorites.

When a location is added to the Favorites list, any time you bring up the search screen on the watch's Map app, the location will appear on the list of Favorite locations.

Getting Directions

The real value of Maps on your watch is not just finding locations and destinations, but getting step by step guides to it. You can use Siri or the Maps app to get locations to a destination while your iPhone is tucked safely away. To do this:

Step 1: Enable Siri. Do this by lifting your arm and using the phrase "Hey Siri," or by pressing and holding the Digital Crown.

Step 2: Ask Siri to find a location or to take you somewhere. For instance, you can say "Take me to the closest Starbucks" or "Directions to the closest hospital?"

Step 3: Locate your desired destination in the Maps app then tap to get driving or walking directions.

Step 4: Tap Start to get step by step directions on the route.

Step 5: Follow the directions your watch provides. To view the steps along with the route on the map you can also swipe left on the screen to see simple readable directions.

If you would like to preview upcoming steps before getting to them, simply use the Digital Crown. If you get off track while following watch directions, the watch automatically recalculates the route as required.

As you drive or walk, the watch provides taptic and audio feedback so you can follow directions without the need to glance at the watch.

When you get to your destination, the directions stop.

Chapter 10: Calendars and Reminders

The march of personalized technology has always been characterized by a synchronized calendar. We all have schedules of different types, and companies have rushed to make sure everyone is covered, from the PalmPilot to the iPhone, the iPad, and now the Apple Watch. What better place to be reminded of an event and activity than on the device you use to check the time – the watch?

Disappointingly, the watch does not have a Reminders app, but you can get alerts from the Reminders app on your iPhone and use Siri to create new reminders on the watch.

Checking Your Schedule

The Calendar app lets you browse upcoming events, and you can of course always use Siri to get specific scheduling information just by asking nicely.

Calendar App

The most straightforward ways to interact with the calendar app, is by tapping the calendar icon on the Home Screen or by asking Siri to open the app. Several watch faces feature a calendar complication that lets you view the next event on your calendar schedule. When you tap on the complication, the calendar app will automatically open.

Alternatively, you can bring up the Calendar glance to see your next appointments.

The default view on the calendar is the Day view. In this view, you can see today's schedule in one scrolling screen, each color-coded depending on the calendars you use on your iPhone. You can switch between days by swiping left or right.

The list view is an alternative to the single Day view. to use the List view, , force-touch the primary Day view screen in the Calendar app then tap the List button.

The Digital Crown makes scrolling events easier and faster. You can alternatively scroll using touch to reveal events up to seven days in the future. In either Day or List view, you can tap an event to get more details about it in a Read-Only view. This means you cannot edit it, but you can read any notes and other information about the event.

You can also see events for the entire month in a Monthly grid on your watch. To see an entire month, tap the date at the top left corner. Note that this just brings a view of the dates — tapping it reverts back to Today and not a specific date you tap.

Siri

You can ask Siri a question directly e.g. "what is my schedule today" or ask it to set an appointment. Feel fry to try asking specific questions about your schedule or setting appointments.

Creating New Events and Reminders

Siri is the only way to create reminders and calendar items on the Apple Watch. Start Siri by raising the watch and saying, "Hey Siri" or by pressing and holding the Digital Crown, then saying something like, "Set an appointment with Jim for tomorrow at 9.30am." If you mention a contact, Siri will ask you to select the contact's email address to send them an invitation. You can then confirm the event and add it to your default calendar.

The same procedure applies to reminders. Ask Siri to set a reminder, either at a specific time in the future or open-ended . You can as well add items to lists in the Reminders app on your iPhone using a phrase like, "Hey Siri, Add carrots to my Groceries list."

When a timed reminder is due, a notification will appear and will include the option to snooze it. When you snooze, the reminder will show up again in 15 minutes. You can tap Completed just below snooze to mark it as done.

Chapter 11: Capture and View Photos

Although the Apple Watch has a small screen, it still has a sufficiently high to show photos. The photos you mark as a favorite on the Photos app on your iPhone or OSX app on your Mac will appear automatically on the watch.

You can also take photos on the Apple Watch — sort of. Using the Camera Remote app on the Apple Watch to control the iPhone's Camera app , the watch can remotely control the iPhone shutter to capture pictures. This is great feature for taking a group shot that you are in, without the need to run to beat the timer or hold the iPhone at arm's length or use a selfie stick.

View Photos

You can view photos on your watch using the Photos app. Scroll through the picture files using the finger or Digital Crown and zoom in by pinching or using the Digital Crown. Swiping the photos pans them and tapping on a thumbnail enlarges it in increments.

When viewing a single photo, the image covers watch's screen. Double tapping on the image enlarges it to show borders.

Selecting Photos to Show

Ordinarily, photos in your Favorites album are copied to the watch, but you can select a different album, including the albums automatically created by the Photos app, e.g. Recently Added or All Photos. To do this:

Step 1: On the Apple Watch app on your iPhone, go to Photos.

Step 2: Tap Synced Album.

Step 3: Select an album from the list then go back to the previous screen.

Step 4: Tap on Photos Limit.

Step 5: Choose the number of photos to transfer.

The watch will store up to 500 photos in its in-built memory, taking up to 75 MB of storage space. If the selected album exceeds the number or size, the most recent photos will be saved on the watch.

Controlling the iPhone Camera

To take photos from the Apple Watch using the iPhone, do the following:

Step 1: On the watch, open the Camera Remote app. This will automatically open the Camera app on your iPhone even if the phone is locked. The app on the watch will display what you should see on the iPhone's camera.

Step 2: Set focus and sample the exposure by tapping the preview area on the watch.

Step 3: To take a single photo immediately, tap the Shutter ⭕ button on the watch. To take a 10 shot burst with a 3 second delay, tap on the Timer 🔘 button.

Step 4: Tap the photo thumbnail on the bottom left corner to review the taken shot. The photos you take will be stored in the Camera Roll album on your iPhone.

Chapter 12: Control Media Remotely

The Apple Watch is a very convenient remote controller because it is always with you, whereas you can be farther from other devices and remote controllers have a habit of vanishing. As long as your iPhone is within Bluetooth range, you can leave it on a table or in your pocket. The same applies to devices on your network: With the watch, you can now pause Apple TV without looking for the remote, or turn up the music on your Mac all the way from across the room. The Remote app on your watch will never get lost between the couch cushions.

The Music app can controls audio playback on your iPhone as well. You can use it when listening to music through the earbuds and the phone is the pocket, or when the iPhone is connected to external speakers. Alternatively, you can leave your iPhone at home and just Sync a Playlist to the Apple Watch and play via your Bluetooth headset.

Pairing the Remote App

Step 1: On the Apple Watch, open the Remote app.

Step 2: Tap the Add Device button on the Devices screen to bring up a four-digit code.

Step 3: Go to iTunes on your computer and click the Apple Watch remote icon when it appears.

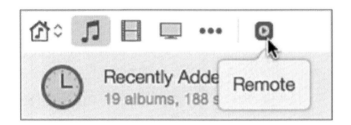

On the Apple TV, navigate to Settings > General > Remotes then select the watch name.

Step 4: Enter the code on the watch to pair the devices.

To remove a paired device, force-touch on the devices screen then tap the Edit button then the X button next to the device to remove. To remove multiple devices at a go, tap the checkmark at the top-right corner of the screen to finish.

Controlling iTunes

Open the Remote app on the watch, then choose a computer running iTunes. You will be able to see the currently playing song or video on the watch. Note that the watch cannot select songs or videos from your iTunes library, you will need to select files to play on the actual computer or using the Remote app on your iPhone.

The media remote control options available on the watch are:

- **Play/Pause** to Start or stop playback.
- **Change tracks** by tapping the Previous or Next buttons.

- **Adjust the volume** by turning the Digital Crown or tapping the plus or minus buttons.
- **AirPlay** by Force-touching the playback screen, then tap the AirPlay  button, and choosing a destination. This is a handy feature when you want to send audio or video playback to an Apple TV or AirPlay-capable speakers.
- **Go back** to the Devices screen by tapping the menu ▤ button.

Controlling Apple TV

Go to the Devices screen on the Remote app on the watch, select your paired Apple TV to use any of these control options:

Select the highlighted option by tapping the screen.

Move the highlight by swiping in the desired direction.

Go up one level by tapping Menu.

Go to the **Apple TV Home screen** by touching and holding the Menu.

Start/Stop by tapping the Play/Pause button.

Fast-forward by swipe right once. A blue triangle to the right of the progress bar will indicate a 1x fast-forward speed. When you swipe right again, it will speed up the fast-forward and two blue triangles will show. You can swipe right again to get the fastest3x speed. Swiping left slows down playback and tapping resumes normal playback.

Rewind follows the same procedure as fast-forward, except that you swipe left.

Playing Music on the iPhone

The Apple Watch Music app controls iPhone audio playback. Open the main Music screen on the watch and navigate to a song then tap it to start playing.

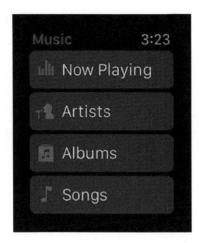

The next song to play will depend on where the previous song was played from. If you play a track from an album, the next songs played will be from that album. You can choose to play all music tracks by force-touching the Music app screen then selecting Play All to add all albums to the playback queue.

While music is playing the available controls on the Now Playing screen include Play/Pause, Skip to play the next or previous track and volume up and down buttons.

You can also access other playback options on the Now Playing screen when you Force-touch the screen. These include Shuffle, Repeat, Source and AirPlay. Source of course lets you select the tracks to play while AirPlay lets you send playback to an enabled device such as Apple TV

While on the watch face, you can access the Music app and controls by bringing up the Music Glance. On the watch face, swipe up to bring glances and swipe right or left to find the Glance with basic music playback controls. You can also toggle the Music app and glance by tapping the song title or artist.

Syncing a Playlist to the Watch

If you want to listen to music without being tethered to the iPhone — for instance when going for a jog — you can simply sync an iTunes playlist to the Apple Watch and play via Bluetooth headphones (Apple Watch cannot play back music directly on its speakers, you must use wireless headphones)

Step 1: Go to the Apple Watch app on you iPhone, then tap the Music item.

Step 2: Tap Synced Playlist.

Step 3: Select a playlist from the list to mark it with a checkmark. You will see "Sync Pending…"

Step 4: Attach the watch to its charger to begin syncing. The watch will only transfer music files when it is charging.

You can manage the amount of storage that synced music occupy on the watch. Open the Music app on the iPhone and tap on Playlist Limit then choose the amount of storage between 100MB and 2.0GB or select the number of songs between 15 and 250 songs.

To start playing synced music, open the Music app on your watch. Force-touch on the Music screen then tap to see a Settings button to pair your speakers if you have not already paired an output device to the watch. When you tap on the Settings > Bluetooth button, you should see nearby discoverable headphones or speakers on the list.

Chapter 13: Using Apple Pay and Passbook

Before you get a chance to use your Apple Watch to purchase something, the idea of paying using a device on your wrist feels almost like a practical joke. But yes, you can. You can walk up to an NFC payment terminal, get your wrist close to the machine and that is it – payment is done. It is an extremely safe, convenient and fast.

Instead of having to handle and swipe cards, enter card numbers or dig up your wallet for cash, with Apple Pay on your Apple Watch, you will just need to get your watch close to the terminal and the numbers stored in the merchant's database will be called up to effect the payment. The Apple Pay app sends a token to the system that will be matched up with the card you set up. Even if a hacker were to intercept the data it would be useless to the attacker.

Passbook is a convenient system for storing items that use barcodes for transactions e.g. loyalty cards flights ticket, movie tickets, baseball game tickets etc.

Setting Up Cards for Apple Pay

Even if you have Apple Pay already set up on your iPhone, you will still need to add cards separately to the watch to use Apple Pay.

Step 1: On your iPhone, open the Apple Watch app then tap Passbook & Apple Pay.

Step 2: Tap on the Add Credit or Debit Card button.

Step 3: Follow the instructions on the screen. These will entail scanning your credit or debit card on the iPhone's camera or manually entering the details on a text area then and accepting terms of service.

Before using Apple Pay, you will also need to complete a verification process, which will be completed by phone call, text or via email depending on the provider.

When the verification process is completed, the added cards will appear in the Apple Watch app as well as on the watch itself. To view them, open the Passbook & Apple Pay app.

If more than one cards are set up, you can select the primary card to use by tapping the Default Card setting and selecting the card you wish to use most.

Using Apple Pay

Step 1: Double-press the side button to launch Apple Pay.

Step 2: If you have multiple cards set up, swipe left or right to find the one you want to use.

Step 3: Extend the watch screen close to the payment reader.

When the payment transaction occurs, you will feel a taptic pulse and a tone alert. The display will read Done when completed.

Passbook

You can set up Passbook items in the Passbook app on your iPhone, using either other apps or email. The passes, when set, will transfer to the watch automatically. To use them, just open the Passbook & Apple Pay app on your watch then tap the item you want to use. You may need to scroll to find the right pass tap to fill the screen.

To rearrange the passes on your iPhone, open the Passbook app on the phone then drag them into place. This screen also lets you delete used and expired passes.

Chapter 14: Exercise with the Apple Watch

For most users and reviewers, the Apple Watch's fitness features are the most useful and the most admired. The watch features sensors that track the heart rate and measure movement and the connection to the iPhone provides accurate acceleration data and location information. The watch has two apps included that make the watch ideal for fitness enthusiasts and generally active individuals. These are the features that really make watch a great part of user's lives.

The Activity app quietly reminds the user to stay active while the Workout app is a personal trainer that pushes them to the next goal. The Workout app offers several common exercises including running, cycling and walking) while tracking your performance during exercise.

Activity Tracking

Open the Activity app right from the Home screen, on the Activity glance, or from the Activity complication on some faces to see your daily activity progress.

You see three rings, each representing one activity accomplished so far in the day. When the rings are complete circles, it means you will have met your day's goals.

- **Calories**: The outermost red ring is a measure of movement and it track calories burned. They are computed based on the amount of activity, your age, sex, and weight.
- **Activity**: The middle green Exercise ring displays recorded "brisk" activity.
- **Standing**: The innermost blue Stand ring keeps track of how often you've stood up and moved around from a sitting position. The objective here is to be in motion for at least a minute every hour.

When you scroll down the activity ring, you will be able to view more details about the exercise progress for each bar, such as a breakdown of when you burned the most calories during the day.

You can adjust your targets in two ways:

- Force-touch the screen and tap the Change Move Goal button to change movement goals. Then, tap the + or − button or use the crown or to set a new calorie goal. When done, tap Update.
- You can wait for the watch to provide its weekly progress report on a Monday which will include a suggestion on new goal and the option to adjust it.

The Activity app on your iPhone will provide the same breakdowns as the Activity app on the watch, but with the added feature to go back and compare with the performance of previous days. On the iPhone, tap on the month at the top of the screen to access a calendar of activity rings. This section also has an Achievements screen where you can see awards you receive for making progress and achieving your activity goals.

Viewing Heart Rate Data

The heart rate sensor on the watch helps determine when you're being active, but is there a way to view the data it collects? Yes.

- Open the Health app on your iPhone then tap the Health Data button.
- Tap on Vitals then Heart Rate.
- Tap on Day, Week, Month or Year to see data from different timeframes. You can also tap Show All Data to see every instance when the heart rate sensor collected data.

Activity Nags

Do you need to be reminded about activities, goals or just want to get rid of the nagging reminders? You can adjust activity reminds using the Apple Watch app the iPhone. The notifications you can get to adjust include:

- **Stand Reminders**: change when to get the "Time to Stand!" notifications on the watch.
- **Progress Updates**: Change when to receive an overview of your activity levels so far. Choose from every 4 hours, 6 hours, 8 hours or None to disable the updates.
- **Goal Completions**: This is a notification when you close an activity ring.
- **Achievements**: When you achieve a set milestone, you will receive a notification. You can view your achievements in the Activity app on the iPhone.
- **Weekly Summary**: Enable or disable the weekly activity summary that is generated arrives on Monday.

Starting the Workout

When it is time to do more than your everyday activity, on the Apple Watch, open the Workout app, or initiate Siri and say, "Start a workout."

Step 1: Select a type of workout, such as Outdoor Walk or Indoor Cycle.

Step 2: Swipe left or right to set the type of goals: time, calories burned, or distance covered (where applicable). You can also tap Open to exercise tracking without setting a goal.

Step 3: Tap Start to begin tracking after a 3-second countdown. The app will begin recording time, pace, heart rate, and distance with the help of the iPhone's GPS capability.

During the exercise session, the Workout app will remain active overriding the Activate on Wrist Raise setting on the watch. You can change the data displayed on the screen by swiping left and right. These include current speed, elapsed time, calories, heart rate and distance, depending on the type of exercise.

When you complete your workout, force-touch the screen, then tap the End button or swipe right to the first screen to complete tracking. You can also swipe to this screen to pause the workout or take a break.

Scroll down to review your workout summary before exiting the app. If you wish, you can save the results to the Activity app on your iPhone by tapping the Save button. Later, you can see the saved records at the bottom of the day's activity details.

Calibrating the Sensors

One of the best activity features of the Apple Watch is that it learns — though in a sort of a limited way — more on how you walk or run. By knowing your average stride, the watch can better estimate your calorie burns, and also track workout distances and paces better even when your iPhone isn't close or when you are indoors. The key here is the watch's accelerometer, which detects steps.

Accelerometer calibration is required the first time you run the Workout app, but you can also calibrate it as you continue to exercise. Follow the directions in the <u>Apple support document</u> you got with the watch to calibrate it by walking or running for 20 minutes in a flat area outdoors where the phone will have a clear GPS signal.

For better results, it is advisable that you repeat the 20-minute walk or run at different but consistent speeds. When it is calibrated, the Apple Watch can be pretty accurate, even without the iPhone.

Andrew McKinnon

Chapter 15: Caring for Your Apple Watch

While it may seem like a sharp-looking timepiece, the Apple Watch is actually a piece of highly complicated and technical electronics. Taking good care of it involves a lot more than just a polish here and there.

Recharging the watch

As I mentioned earlier in this book, expect to recharge the battery every day if you are an average user, and you do this by connecting the included magnetic charging cable to the back of the watch. The charging system should automatically snap to its place.

If you want to conserve the battery life of your watch, it is advisable that you manually put the watch into Power Reserve mode when not actively using it. This shuts down everything except a minimal digital time readout. To do this, from the Battery glance, tap on the Power Reserve button or, simply by pressing and holding the side button until you see the Power Reserve slider on the screen, then sliding it to activate.

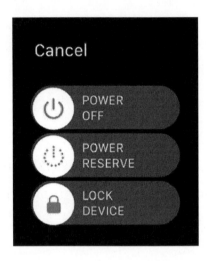

To return to the normal battery mode from the power reserve mode, press and hold the side button again to restart the watch.

Restarting the watch

If the Apple Watch is not behaving as it should, it may be helpful to power it off and then back on. To do this, press and hold the side button on the watch to access the power controls menu then and slide the Power Off slider. You should then wait a few seconds then press and hold the side button until Apple logo appears.

In some cases, if the watch freezes and you need to force-reset it, press and hold both the Digital Crown and the side button until the Apple logo appears.

Resetting the Watch

If something seems exceptionally whacky with your watch, or if you are going to give it away or sell it, you will want to reset it to its factory settings first.

The data on the watch including watch face settings, saved Apple Pay details and media is automatically backed up to the paired iPhone and will not be forever lost when you reset the watch. You should however make sure the iPhone is backed up to a computer via iTunes or to the iCloud before you reset the watch.

To reset your watch, you will essentially be unpairing it from the companion iPhone, which will actually perform a final backup of the data before the watch is wiped clean. Here is how to do it:

Step 1: Go to the Apple Watch app on the paired iPhone, then tap Apple Watch.

Step 2: Tap Unpair Apple Watch, then confirm the action.

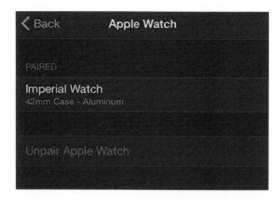

After a few minutes, the Apple Watch will be restored to its factory default state.

If you need to reset the watch but the iPhone isn't available, go to the Settings app on the watch then to General > Reset and tap on the Erase All Content and Settings button. The watch will be ready for the next owner, or to be restored.

Restoring the Watch

After the watch is reset, you can pair it with your iPhone again.

Step 1: Select your language then tap the Start Pairing button.

Step 2: Go to the Apple Watch app on the iPhone and tap Start Pairing button then scan the animated cloud on the watch. This process is the exact same one you used to pair the watch with your iPhone at the beginning.

Step 3: When the watch is paired, tap on Restore from Backup button.

Step 4: Select the watch's latest backup.

Step 5: Agree to Apple's terms and conditions.

Step 6: Enter your Apple ID and password to enable Apple Pay cards, Digital Touch, and other features. If you have enabled two-step verification, you will be required to enter the code that is sent to you.

After the remaining setup steps including enabling Location Services, etc, the iPhone will sync data back to your Apple Watch. In no time, the watch will be ready for use.

Updating Software on the Watch

When Apple releases a new update to the watch's operating system, you should receive a notification. You can update your watch by opening the Apple Watch app on your iPhone then navigating to General > Software Update.

Cleaning the Watch

Apple's advice on cleaning your Apple Watch is straightforward commonsense: wipe your watch with a lint-free, nonabrasive cloth, wetting it with fresh clean water if required. Apple also recommends that you dry the watch and its band after exercise every, which is important for its durability as well as your hygiene.

The watch is designed to be water resistant, so it can wear it into the shower. However, Apple does not recommend submerging it in water despite its IPX7 rating. The rating means that the watch can withstand up to 1 meter of submersion in water for up to 30 minutes).

If the Digital Crown is not turning or becomes unresponsive, it is advised that you run it under "lightly running, fresh, warm water from a faucet for about 10 to 15 seconds."

What to Expect from watchOS 2

During the 2015 WWDC, Apple made a big announcement regarding the Apple watch. Finally, the company confirmed that it was developing the second generation watchOS for the Apple Watch, and it will pack your watch with a ton of new features and capabilities. When the WatchOS 2 update arrives, it will finally have the native app capability meaning that the watch will be able to run its own apps rather than be constantly be tethered to the iPhone Native apps are expected to run faster and have greater capabilities than tethered apps.

The wrist device is expected to have a ton of new and refreshed features to enrich the user's experience. If you already have the Apple Watch, you may not need to buy a new one to take advantage of these features. You will just need to upgrade the OS to get access to them. Here is a breakdown of the confirmed features and specs of the Apple Watch running watchOS 2:

1. Native Apps

When watchOS2 arrives, one of the most anticipated features is that it will allow developers to create apps that can be installed on the watch, meaning that they can run without the iPhone being connected. Apple has confirmed that native apps have been granted greater control and access to watch functions such as the speaker, microphone and inbuilt sensors like the accelerometer and network access. The Digital Crown and Taptic feedback have also been opened so that applications can make use of all inputs, controls and feedback tools of the watch. With watchOS 2, for instance, you may be able to adjust the temperature in your car just by rotating the digital crown.

2. New Faces and Complications

The faces on the current version of watchOS are very limited. The new operating system update is said to expand the Photo Face so you can make a watch face from your favorite pictures and even rotate through a selection of your photos if you wish so. One of the most anticipated features is the time-lapse video covering five cities – London, Hong Kong, Mack Lake, Shanghai and New York that lapses over a period of 24 hours to create a unique new face for the watch. Furthermore, the new complications will also be able to display information from third party applications such as Twitter feeds.

3. Time Travel

Time Travel is a new Apple feature that lets you speed forward in time just by rotating the Digital Crown on your watch. This feature will enable you to quickly check what is coming up in the day or hours ahead and even show you how your battery is going to perform based on current usage. With this information, you will know when you need to charge. This feature is a lot like Pebble's Time Key feature, but we expect it to have additional features and capabilities.

4. Nightstand Mode

Nightstand Mode is a new feature on the Apple Watch that flips the time horizontally so you can keep the watch showing time while on its side and charging the whole night. This new feature also offers a chirpy alarm functionality to wake you up in the morning and you can snooze it with the Digital Crown. You can also choose whether to keep the screen on the whole night or keep it off.

5. Communication

The current watchOS only lets you have 12 "friends' on your watch. Well, with the update, you will be able to have multiple circles of friends categorized into Work, Friends and even Friends with Apple Watch. Apple has also added multiple colors to the Digital Touch for deeper doodles and you will be able to respond to your emails right from your watch. Another great feature added to the watch is FaceTime audio. However, this new addition will only let you make brief vine-like videos and audio, but there is a great possibility the functionality will be expanded.

6. Health, Fitness, Siri and Maps

With the watchOS 2 update, workouts will be a lot more useful and HealthKit will come with a lot more metrics including the display of real-time heart rate streaming on your wrist. The Fitness app will now be able to work with third-party applications and you can add your own workout schedules right from the Apple Metrics. Siri is also sharper on the update as you will be able to control the HomeKit devices and even ask Siri to read data from glances, get directions and seamlessly set reminders. The Maps app features new transit directions the include train, bus and subway schedules for select cities around the world.

7. Wallet and Apple Pay

Apple Watch now lets you make payments using your card right form your watch and even claim rewards using the Passbook. With the update to come, the Passbook is changing to the Wallet, and will of course have expanded capabilities and features. The new application will bring more features like those available in the iOS9 wallet to your wrist. If you already enjoy the convenience of Apple Pay on your Apple Watch, you will be happy with the watchOS 2 because it will support even more credit cards and reward cards for better and more convenient access.

8. Activation Lock

Apple's new watchOS 2 comes with a new feature called Activation Lock. This is a new security feature that is will require you to use your iCloud Apple ID and password to secure your watch. This feature currently exists in iOS devices.

9. Sensors and Controls

The functionalities of the Digital Crown have been expanded in the watchOS 2 beyond zooming in and out and scrolling. Besides Timetravel, you can also control lights in your house with the Digital Crown with Insteon and a number of other uses. Since you will finally be able to access the microphone, making voice memos on your watch will be a breeze. You will also be able to set different vibrational feedbacks and sounds for different third-party apps as the taptic engine is now open to developers.

10. Tetherless Wi-Fi

Tetherless Wi-Fi is exactly what you think it is, and it is going to make the watch great. When your watch gets the watchOS 2 update, you will be able to connect to Wi-Fi networks without even having your iPhone.

Printed in Great Britain
by Amazon.co.uk, Ltd.,
Marston Gate.